THE INVENTOR'S GUIDE
FOR MEDICAL TECHNOLOGY

FROM YOUR NAPKIN TO THE MARKET

WHAT INNOVATORS NEED TO KNOW

THE INVENTOR'S GUIDE
FOR MEDICAL TECHNOLOGY

FROM YOUR NAPKIN TO THE MARKET

WHAT INNOVATORS NEED TO KNOW

PATRICK KULLMANN, MBA

Two Harbors Press
212 3rd Avenue North, Suite 290
Minneapolis, MN 55401
612.455.2293
www.TwoHarborsPress.com

ISBN-13: 978-1-937293-85-7
LCCN: 2011942364

Distributed by Itasca Books

Cover Design and Typeset by James Arneson

Printed in the United States of America

THIS BOOK is dedicated to my beautiful wife, Pamela, in honor of her selfless service to others, especially to those in need; to my daughter, Bethany, who has grown into a beautiful young woman with a caring and confident leadership style; and to my parents, who loved me.

I want to also recognize and dedicate this book to the men and women who take risks innovating and creating businesses in the health care industry and the expert people who provide the services that make them successful so that we can all have the highest standard of medical care.

CONTENTS

FOREWORD

When Patrick and I first met at the suggestion of my corporate attorney, I thought he was a potential investor in our newly formed medical device company. I guess he thought of me as a potential client interested in one or more of his multiple business and marketing services. As we sat over breakfast, we made our respective sales presentations to each other but finally realized the error of our perceptions.

That initial meeting, despite its miscues, led to a number of other meetings, usually at local coffee shops, that resulted in our focused efforts at starting a new company devoted to helping health care workers mature their problem-solving ideas into clinically successful products and services.

It did not take me long to realize, after my conversations with Patrick, that starting a medical device company based on my inventions need not follow a haphazard sequence but could actually consist of a pathway that can, more often than not, lead to a successful conclusion. From personal experience, I can tell you that learning in the trenches, while trying to avoid the land mines that surely await the uninitiated, can be very trying, both emotionally and financially. It's for this reason—to help those physicians, nurses, and other providers who have great, clinically relevant ideas but lack the knowledge to carry them forward—that this book was created. It will best serve those whose egos do not interfere with the realities of business as presented by Mr. Kullmann. His extensive experience with the industry will rapidly be apparent in his comments that end each chapter. Certainly, he knows of what he speaks.

If I'd had access to his experience before I started my first medical device company, I would have chosen my management staff more carefully, would have saved three to four years in achieving a more efficient iteration of my product, and would have fashioned a more rapid route to market acceptance. Trial and error is not the best teacher when it comes to producing a medical device that needs to pass an FDA or a Notified Body audit when seeking a 510(k) exempt status or a CE mark, respectively, for sale outside the United States.

Mr. Kullmann's significant contribution to the inventor/innovator/entrepreneur is that he has taken his lengthy personal experience in the medical technology industry and distilled it down into a sequence of steps that promises to lessen the risk of failure and to heighten the chance that the practice of medicine will benefit from the genius that dwells within those that practice the healing arts. Perhaps that first meeting between Pat and me was not a miscue after all but rather a first step in my business education that would allow me to create my third medical device company.

Thank you, Patrick.

Leonard Schultz, MD
General Surgeon
Founder and Chairman, Nascent Surgical, LLC
Minneapolis, MN

ACKNOWLEDGMENTS

I would like to thank the following medical technology service professionals for their contributions, advice, guidance, and counsel in writing this book. They are people who are key thought leaders in their areas of medical technology expertise and are responsible for contributing to the success of this book and of many medical technology companies around the globe.

A special thanks to:

David F. Benusa, CPA, MBT, CEO
Froehling Anderson, a CPA firm

Edward Black, President
Reimbursement Strategies, LLC

John Deedrick
Managing Director, Linn Grove Ventures LLC
Venture capitalist

Mark DuVal, JD, President DuVal & Associates, PA
Food, drug, and medical device law

Timothy Hannon, MD, MBA
President & CEO, Strategic Healthcare Group LLC

Kathleen Leith, MBA, health care consultant
CG3 Consulting LLC

Kermit J. Nash, JD, principle
Entrepreneurial Services, Gray Plant Mooty

Paula J. Norbom, President
Talencio, LLC

Karen Nordahl, MS, health care economics consultant
Founder and Principal Advisor, Running Rock, LLC

J. Robert Paulson Jr., JD, MBA
Medical technology entrepreneur

Doug Pletcher, PE, MBA, health care consultant
CG3 Consulting LLC

Leonard Schultz, MD
General Surgeon, Founder and Chairman, Nascent Surgical, LLC

Ann Quinlan-Smith, MBA, President
Alquest, a NAMSA company, clinical research organization

Benjamin A. Tramm, JD, BA, BSEE
Merchant & Gould, an intellectual law firm

W. Robert Worrell, Founder
Worrell, Inc., a global product design and development firm

INTRODUCTION

Late in 1992, our company was poised to be purchased by one of the largest medical technology companies in the world, which had acted as a strategic investor. We had developed a technology for measuring continuous cardiac output, or CCO, an important hemodynamic parameter for the medical treatment of critically ill patients. Until this time, several inventors had tried and failed to provide accurate cardiac output measurements in a continuous manner, but we were successful. The inventor, physician, and founder of the start-up company was a wise man with medical technology experience as well as clinical experience as an anesthesiologist. Most important, he had surrounded himself with highly skilled and experienced people.

The strategic investor had made an investment in the company and was granted a seat on the board of directors of the company with the "first right of refusal" to acquire the company if it accomplished certain performance milestones. Well, the time had arrived, and we sold the company to one of the largest medical technology companies in the world for $60 million.

Later in 2001 we positioned a second company for acquisition that was founded by an engineer in 1989. We grew it significantly, and it was acquired for $60 million by an equity fund. The company was renamed and continued its acquisition of other start-up medical technologies and grew to more than $500 million in annual sales and went public via an initial public offering (IPO). In late 2010, this company was purchased by a strategic buyer for $2.6 billion.

Both companies made significant contributions to the treatment of patients and to the building of wealth for the inventors, founders, shareholders, and employees, as well as for local, state, and federal government, which benefited from the tax receipts that it had generated.

The idea for this book came to me after starting CG3 Consulting LLC. During my twenty-eight years in the medical technology field, I have worked for such notable companies as Johnson & Johnson, Medtronic, Boston Scientific, Baxter International, and four start-up medical technology companies in ten medical specialties. I have worked with many new medical technologies, and I learned over the years that the roles of physicians and other health care providers, as well as engineers, and scientists as innovators, are paramount in innovating medical technologies for the continued fight against the diseases that plague mankind. It is a fact that more medical technologies are conceptualized by physicians and other health care providers each year than any other innovation segment. Scientists develop the compounds for drugs, but health care providers—hands down—create the most concepts for devices. Providers see the unmet needs in technology and patient treatment in their everyday practices and thus seek solutions. They are problem solvers.

I realized that there was a gap, however, in the development process for these inventors. Most innovators are first-time inventors, both in the United States and abroad, who have not had the experience of taking a product or service to the market successfully, especially in the U.S. market. Many inventors have not had the experience of understanding what I refer to as the process of taking their idea from concept to market. This is why I think that the contents of the book will be relevant to the reader's needs.

The purpose of this book is to provide an introductory overview of some of the things that an inventor needs to know and should consider. I had considered writing a lengthier book on this subject, but I felt that a shorter version that was a "quick read" with all of the essentials would be more useful to the reader, a book that would offer practical advice based on reality and my experience and the experience of other experts, not simply theory. I was not convinced that people want to read another four-hundred-page textbook.

It is this author's hope that this book will provide a basic foundational guide for the inventor and thus increase the probability of success. Although following the guidance from this book will not guarantee success, it will provide information on the general pathway to success in medical technology commercialization.

The intent of this book is not to fully equip inventors and entrepreneurs to act in a solo role in commercializing their idea but rather to advise them on some of the major areas that must be considered in order to be successful. Readers, if they are honest with themselves, will realize that it requires much more than reading and re-reading this book or other informational materials to be successful in the medical technology arena—they need the assistance of experts.

Inventors and entrepreneurs have very different skills. *Webster's* defines an entrepreneur as the one who organizes, manages, and assumes the risks of a business or enterprise. An inventor is, by definition, one who produces something for the first time through the use of the imagination or ingenious thinking and experiment. Sometimes the inventor will become the entrepreneur; sometimes it will be a different person with entrepreneurial skills. The wise inventor and founder will understand the difference. This book is directed to both of these people. In the book we will use both terms—the principles apply to both.

For the purposes of this book, we will use the following definition of medical technology: Medical technology includes a wide scope of health care products and services that, in one form or another, are used to serve, diagnose, monitor, or treat diseases or medical conditions. We will focus primarily in this book on medical devices for illustration purposes; the principles of this book, however, can apply to health care services, software, social media, and other emerging health care-related products and services.

There are many opportunities to create treatment and diagnostic solutions for the diseases that impact us in our daily lives, as well as to concurrently build wealth and economic opportunity for others who are consultants and employees. The inventor and entrepreneur will face significant opportunities and challenges, the first being a realization of what he or she does *not* know and what needs to be considered and learned in order for an idea to become successful. The genesis of a medical technology start-up comes from courage—courage to take the risk, courage to make a difference in the lives of patients.

This book will outline many of the key considerations for success for both the inventor and entrepreneur, based on my experience. It offers a basic introduction to the process of taking a medical technology idea from concept to reality.

Enjoy the book, and best wishes for future success.

Patrick Kullmann, MBA
Founder and Lead Strategist, CG3 Consulting LLC

My Advice:
Inventors should not attempt to commercialize a medical technology invention on their own; they need the assistance of experienced medical technology professionals.

CHAPTER 1

THE INVENTION

Necessity is the mother of invention. We have all heard this, and it is definitely true. Without an unmet need or want, buyers are not motivated to purchase anything, and medicine is no exception. The most successful inventions result from cost-effective solutions to recognized clinical problems that the market wants solved. The relationship between success and failure of the start-up medical technology company cannot be dependent on anything else as important as this.

Innovation is a wonderful gift to a free society. Men and women who dream of ways to improve the lives of others and have the opportunity to create wealth and employment opportunities where they work and live are commendable.

Take the case of cardiovascular medicine and the way that innovation has impacted the improvement of patient care, as well as the industries that it has created. Ischemic heart disease once was treated by primitive drugs and by placing the patient in a chair to rest during episodes of angina. Open-heart bypass surgery, followed by coronary angioplasty and stenting, became follow-on advances for treatment. Today, drug-eluting stents and statin drugs have become an effective line of defense for ischemic heart disease. Many of these devices were conceived by inventors that include physicians, engineers, and other scientists.

Inventors have to think objectively. It is common for them to become emotionally attached to their invention; after all, they have invested time, money, and energy into their "baby." Because of this attachment, many inventors only see opportunity and not obstacles. More important, they may not appreciate the skills, hard work, time, and financial resources that will be needed to fully commercialize their product. Many inventors believe that if they just have the opportunity to demonstrate their product for a major company audience that the company will immediately appreciate its value and acquire it right away. Companies rarely acquire an early technology until "proof of concept" and "proof of principle" is clearly established. We will discuss these important concepts later in this book.

My Advice:

Taking a product from concept to market is very hard work, with significant risk. It is a process that you, as the inventor, should not attempt to do on your own. My first piece of advice is to seek outside professional assistance before you begin—and before you spend your first dollar. It will save you much in financial expense and provide you fewer sleepless nights. Fall out of love with your invention, and look at it very critically—I guarantee that others will.

Is It an Idea, a Product, or a Company?

Is a product idea a product … or just an idea? It is important for you, the inventor, to understand, grasp, and accept this question. After all, it is your invention, and it can be very personal. I will tell you something straight up—*I advise all of my early-stage clients that if they are going to fail, they should fail fast and early; it is less emotionally and financially painful.* What I mean by failing early is that we need to determine the probability of success for your product idea earlier vs. later. It is better to know this probability as early as possible, before significant resources, effort, and time are invested. I realize that this is direct, and discussing this so early in this book sounds counterintuitive, but it is very important. *We need to know what we need to know, and we need to know what we don't know.* If we know the truth early and understand it, we can manage it appropriately.

Inventors will often attempt to form a company around their product idea prematurely. Now, I'm not saying that you should not form a legal corporation or other entity under consultation with legal counsel and your accounting representative to determine the best corporate structure for your situation. What I am saying is that a single product does not usually make a company; it depends on the size of the market. A single product idea can, in some cases, make a company, with its variations allowing for a later product line. Multiple products are not needed to make a company successful, although that is often what happens. If the inventor has a technology that is a foundational base for a family of products, then the discussion moves from *idea*, to *product,* to *company*.

An idea is a concept of what a product should be, what it should look like, and how it should be used. It has not taken shape yet, even as a model of the product. This idea does not become a product until the steps in this book are accomplished, resulting in a physical device or service that serves patients. A company is a sustainable entity that offers several products in a family portfolio offered by the company. Often, inventors have a vision for an exit that they think will result from a major corporation's acquiring their company, when really, an acquirer is interested in the product, base technology, or patent practice rights, depending on the development stage of the product.

My Advice:
Face the truth about your invention: will it work and will people buy it? Be realistic.

CHAPTER 2

THE MARKET OPPORTUNITY

Determining the market size and opportunity is the most important item in the early stage due-diligence process; it is commonly misunderstood. Early stage due diligence is a different process from "deal" due diligence, which is accomplished by an acquiring or investing party when considering investing or acquiring the company or product. Early stage due diligence is an evaluation process that occurs at the beginning and in the early stages of your company to better understand if your idea will work in order to manage risk and maximize opportunities.

Understanding a few introductory terms, such as the *total market*, *potential market*, and *available market*, will help you in planning for a successful enterprise. The inventor should consider the incidence and prevalence of the targeted disease state where the product will be used on an annualized basis. Prevalence measures how much of some disease or condition there is in a population at a particular point in time. Incidence measures the rate of occurrence of new cases of a disease or condition.

Total Market

Using the example of the United States, its population of 311 million people is a representation of the total market. At the beginning you should include all people who could receive the therapy or diagnostic service of the product over a life span. Next, you'll need to understand the potential market within the United States. For this illustration, we will consider a device that treats ischemic heart disease as a case study.

Potential Market

We know that ischemic heart disease primarily strikes patients older than 35 years of age, so we would consider the potential market in the United States as all people over the age of 35.

Available Market

Now, the available market is that sub-segment of the potential market who potentially would be treated with the inventor's device. In this case, we will use an example of an interventional cardiology product for primary coronary interventions (PCI), or coronary angioplasty. This available market would then include the patients who have ischemic heart disease who are treated with PCI, not with bypass surgery or medical management. We must arrive at this market-size data to accurately assess the inventor's product, if it were used in place of PCI.

Understanding the diagnostic and treatment continuum is important to your success. Will the introduction of your product or service change or disrupt the medical specialist who traditionally treats patients with a particular disease? Many times, a new and disruptive technology creates new winning and losing users of technology in the treatment of patients. Using our illustration with PCI and cardiac surgery, the introduction of PCI in 1979 and over time was considered a disruptive technology to traditional cardiac bypass surgery. Statin drugs were disruptive as well.

The market is also segmented into early adopters of technology, mainstream users, the largest part of the available market, and late-stage users. Companies are successful because of "mainstream" buyers, not early adopters, so we need to quantify the available market size of mainstream buyers. Early adopters are useful as key opinion leaders (KOLs) because they provide roles as supporting authors of peer-reviewed articles and supporting your invention by acting as speakers at medical meetings, but mainstream buyers drive growth and the value of your company.

My Advice:
It is available market and mainstream buyers and the time to adoption that counts. Everything else is an overinflated concept.

CHAPTER 3

REGULATORY

Regulatory issues are some of the most important areas that inventors and entrepreneurs must consider in evaluating the opportunity for commercializing their products. The timeline and financial resources needed should be an important consideration in the inventor's early due-diligence process before moving ahead too quickly. Shorter pathways for FDA approval usually require fewer financial resources than longer pathways. Investors are very sensitive to the costs and timeline in achieving the milestone for U.S. approval of a medical technology. Here are a few things that you need to know:

United States

The FDA, under the Food, Drug, and Cosmetic Act, classifies medical devices into three classes, based upon their risk: Class I—low risk (tongue depressors, wheelchairs); Class II—moderate risk (ablation devices, stents, many orthopedic devices); Class III—high risk (implantable devices like defibrillators, neuromodulation devices, drug delivery pumps). The three basic pathways to the market are the 510(k) pre-market notification, the pre-market approval (PMA), and the de novo provisions of the Food, Drug, and Cosmetic Act. What follows is a brief primer of these provisions.

The Pre-Market Approval Process (PMA)

The PMA is the highest risk pathway, known as Class III. For a PMA the sponsor must demonstrate that the device is safe and effective in an absolute sense, not in a borrowed sense, as with a 510(k). Unlike the 510(k), the PMA pathway is used for novel devices and/or those that involve much more risk than a Class II device. The standard is to demonstrate "reasonable assurance of safety and effectiveness." PMA devices are Class III devices that have the highest risk and include devices like implantable pacemakers, deep-brain stimulation devices, and other devices through which risk of mortality and life impairment are high. PMAs require a great deal of human clinical substantiation and many years of study and FDA review before they are approved. The PMA is the most time-consuming and expensive pathway.

The 510(k)

The 510(k) pathway is reserved for moderate risk, Class II devices. *Most medical devices today are cleared under the 510(k) provisions.* With a 510(k), all the sponsor needs to show is that the device is "substantially equivalent" to a predicate device that has been on the market, known as a "pre-amendments device." The rationale is that there are devices that have been safely on the market and a sponsor need not reprove the safety of the new device in an absolute sense, as with a PMA. Instead, the sponsor must show the device is as safe and effective as the predicate to which it claims substantial equivalence. A PMA device cannot serve as a predicate; only another 510(k) device can serve. The 510(k) process is not technically an FDA *approval*; it is a *clearance*, because the sponsor is simply notifying the FDA that the product is coming to market. The FDA has ninety days to determine whether the product is or is not substantially equivalent. It will then send either a Substantially Equivalent (SE) letter to the sponsor if the

FDA agrees, or will give the sponsor a Not Substantially Equivalent (NSE) letter if the FDA disagrees. If the sponsor receives an NSE letter, the product cannot be marketed, and the device is automatically reclassified, by virtue of the statute, into a Class III device (see de novo discussion below).

The focus of the 510(k) is on the comparison between the predicate and the new device. The questions to be answered are as follows:

➲ Does it have the same intended use?

➲ Does it have the same technological characteristics?

➲ If it doesn't have the same technological characteristics, do the new features raise any unanswered questions of safety and effectiveness?

The FDA essentially borrows or draws from the knowledge it has of the predicate device so that data is not needlessly regenerated when the medical device to which it is being compared involves well-established technology. The 510(k) system embodies the idea of administrative efficiency. To state that obtaining a 510(k) is dramatically simpler than obtaining a PMA is bit of a misconception, because 510(k)s today require a lot of substantiation, often look like a PMA, and get a significant amount of FDA review scrutiny before coming to market. Today's 510(k) is often called "PMA-lite." Historically, the FDA did not typically require a company to submit clinical trials for a 510(k) to establish the safety and effectiveness of the device and to obtain FDA review and approval before coming to the market, as with a PMA. Today, the FDA often requires clinical trials for a 510(k), but they will not be as extensive and expensive as for a PMA.

The de novo

The final category for an approval is a de novo. When a company receives a Not Substantially Equivalent (NSE) letter from the FDA after it files for a 510(k) clearance, the company's device is, by operation of the statute, automatically classified as a Class III device, for which a PMA must be filed. The statute allows a company, within thirty days of the NSE letter, to file for a de novo approval, arguing that its device, while it may not have a predicate, is a Class II risk device and should be approved as such. To reiterate, for a de novo, the device has no predicate and therefore cannot obtain a 510(k) substantial equivalence determination. But there is recognition by the FDA that some devices should not automatically be classified as a Class III device because it presents a lower level of risk than is found in a Class I or II device. For a de novo, the sponsor must show the device is safe and effective, as with a PMA, but the burden of proof is supposed to be lower because the risk of the device is lower. The FDA takes into consideration any features, technologies, mechanisms of action, and materials that are known to exist with other devices. As with a 510(k), the FDA draws upon a vast body of institutional knowledge and experience with similar devices and technological features from within that therapeutic segment and others. To do so allows the agency to take administrative notice of things that are known and to focus on what is unknown about the safety or effectiveness of a device. This analytical framework and mind-set saves precious agency and industry resources in the process.

Europe

A medical technology will require CE mark certification (approval) in order to be commercially distributed in various countries outside the United States. CE mark certification of the device will consist of working with an accredited European notified body to determine the appropriate documentation required to support certification in accordance with the medical device directive. Today, many inventors and investors must gather clinical data internationally and prove concept and principle, due to the timelines and unpredictability of the FDA.

My Advice:

Understand the regulatory pathway for your product—the time required for clearance and the associated costs. A long and costly regulatory and clinical pathway can quickly kill some product ideas.

CHAPTER 4

REIMBURSEMENT

Reimbursement issues, such as those of the regulatory kind, are critical to the success of a medical technology start-up and should be considered early in the process. If reimbursement does not exist, or it is not realistically obtainable, or it is cost-prohibitive to obtain, then the project of developing a technology might be in jeopardy. We need to know this early in the process.

For any new product introduced into the U.S. health care system, reimbursement is essential to commercial success. In an era of ever-tightening budget constraints, adequate reimbursement for clinical procedures and new technology can be as important as securing FDA clearance. New technology that is judged to be "safe and effective" (the FDA standard) may not be considered "reasonable and necessary" (the Centers for Medicare and Medicaid Services (CMS) standard) for treatment of specific medical conditions.

Key Considerations

➲ Coding: Is there a procedure code that adequately describes the technology or service?

➲ Coverage: Do government and private health insurers provide coverage, or do they consider it experimental or investigative?

➲ Payment: What is the payment methodology for the technology, and is the resultant payment sufficient for adoption by providers?

There are controlling processes that make these all separate, distinguishable challenges that must be successfully navigated for new technology. Not all services that have a code are covered, and not all covered services are paid adequately. A successful reimbursement strategy identifies these barriers and creates solutions to resolve them.

Health care coverage for most Americans is provided through private insurance plans, Medicare, or Medicaid. Through the Medicare program (administered through CMS), the federal government is the single largest purchaser of hospital and physician services. However, CMS delegates all claims-processing and most coverage policy development to eleven Medicare administrative contractors (such as Noridian Administrative Services, Trailblazer Health Enterprises, National Government Services, and Palmetto Government Benefit Administrators) who administer the program through fifteen regional contracts. As a result, new technology and services in one Medicare jurisdiction may not be covered in another, because coverage decisions are mostly made by the regional contractors under broad guidance from CMS.

Medicare coverage and payment policies have a significant influence on the commercial opportunity and adoption of medical technology in the market. Yet private insurers serve different customers. They are typically younger and healthier, and their premiums are mostly paid by employers. If a product does not have reimbursement, it will not be successful in the U.S. or non-U.S. health care systems. Outside of the United States, reimbursement is more complicated and is usually considered on a country-by-country basis.

My Advice:
The cure for death is a wonderful therapeutic product, but without reimbursement it is not a business.

CHAPTER 5
RESEARCH AND DEVELOPMENT

Proof of Concept

Frequently, inventors look to an engineering or design firm to develop a "proof of concept" model, or working prototype. Developing a prototype can be expensive, so it is very important to make sure you have "freedom to operate"—that is, that your idea does not violate or infringe on other patents—before making this investment. Medical technologies are frequently complicated from a mechanical, electronic, or materials standpoint. It is important to understand that creating a functioning prototype necessarily includes engineering documentation to create the components and system design for assembling the device. This could include electronics, mechanical systems, and software. These proof-of-concept prototypes are usually not created to be attractive and may not actually be designed to work in the environment for which they were conceived. Such a bench-top prototype might only demonstrate the key function or technology that displays the patentable feature or benefit, without consideration of usability, aesthetics, or cost-effectiveness, in order to keep initial costs down.

Investors want to know if the idea is technologically feasible, and such prototypes are a way to demonstrate this. It may not be necessary to have what is called a "quality system" to develop a proof-of-concept model, but if you are going to commercialize a new medical technology, the firm you hire will need to have one. A quality system is an FDA-recognized process that makes sure important steps are implemented and documented (design history) to ensure safety for patients as well as health care staff. This is a *must* for developing medical technologies to ensure you are compliant with FDA guidelines and to help in obtaining 510(k) clearance or PMA approvals to market the product.

If your proof-of-concept prototype does its job and you get funding, your funding level will be dependent on whether more research and development-type work is needed or if you are ready to commercialize. It is very important to understand the difference between R&D work and commercialization, as the estimates for these two activities are dramatically different.

Commercialization assumes the technology has been proven, and it is now a matter of packaging it for the three attributes mentioned earlier. Quotes for this work can sometimes be capped with a "not to exceed" figure if the steps of the process are clear. If commercialization activities are interrupted by technology failures or oversights not worked out in the R&D stage, the costs will escalate, as will timelines. Estimating costs for "new to world" concepts or R&D technology is difficult.

As you contemplate commercialization you also might need to determine the type of firm to help take you forward. There is a difference between engineering firms and design firms. Typically, engineering firms will focus on the functional aspects of the device, making it reliable, durable, and capable of being manufactured cost-effectively—the basics. Design firms typically will be focused more on the value-add aspects of product development beyond the basics, such as customer research, to define what the customer wants or how the new device might impact either the patient or health care workers' behavior, efficiency, or adoption of the device—the experience. Design firms will pay more attention to the human factors or ergonomics of the device, including screen interface, which has now become a very significant focus of the FDA. While functionality and the unique new benefit the device is intended to deliver is important, failure to understand customer needs and device interface can impede the delivery of the technology or functional benefit and lead to failed adoption. Execution is still everything. Some firms will have both engineering and design, but usually there will be one discipline that is stronger than the other, so it is important to determine if your intended market will reward you more for basics or value-add.

R & D Tax Credit

For many start-ups, research and development may help fund the company during the early years and support the growth during mature stages of the business. The R&D tax credit may reduce past, current, and future years' tax liabilities, creating an immediate source of cash. The credit differs from a deduction in that it is an actual dollar-for-dollar offset against taxes owed or paid and is a source of permanent tax savings that can be claimed for all open tax years.

Your company may qualify for the R&D tax credit if you have invested time, money, and resources toward:

- Developing new products or improving existing products
- Developing new materials
- Building and testing prototypes and models
- Developing new or improved software applications
- Testing new concepts
- Developing or improving manufacturing processes
- Experimentation and more

The documentation to support the credits and computation can be complex but the benefit may be substantial, generating tax credits in 4–7 percent of total qualified research expenditures. Most qualifying expenditures result from the wages paid to employees who participate in qualifying activities. It's important to note that if credits cannot be used in a particular year, they are carried back one year and then carried forward up to twenty years.

If your company has been engaged in qualifying research activities for the last several years, you may be eligible to retroactively claim R&D tax credits.

My Advice:
It takes just as much time to develop the wrong product as the right one. Get the customer specifications down from the beginning. Remember, it is more about what mainstream buyers think than what you think.

CHAPTER 6

PROOF OF CONCEPT—PROOF OF PRINCIPLE

irst-time inventors and entrepreneurs usually do not have a fundamental understanding of "proof of concept" and "proof of principle" of their invention. In validating the product idea we must first have some way of demonstrating, in an unbiased way, what the proof of concept is for the product. Simply put, will it do what we think it needs to do? Will it treat the designated disease or diagnosis it as we plan? Does it work? Photographs, video animation, patent drawings, and other early-stage models will help in this validation. It is a standard requirement to have bench, animal, possibly limited human, or other critical testing data to prove early non-human proof of concept.

Proof of principle requires that we look at the product idea in a different way. Is there a market need? What are the sizes of the total, potential and available markets? And most important, will customers buy it not only once but adopt it for routine use at a price that is competitive and profitable? *If we cannot make a profit, then we have a hobby, not a business.*

We need to understand if the mainstream buyers will adopt the product—products and companies cannot be successful with the purchases of early adopters alone. Other things to consider are the status of the patent and the financial and human-skill resources required to commercialize the product or company. Inventors can raise initial or substantive rounds of funding by established proof of concept *and* principle. The acquiring companies usually want this, plus clinical data that demonstrates efficacy.

My Advice:
Proof of concept and proof of principle are the most important elements of success. Without them, you do not have a business; you have a hobby.

CHAPTER 7
CLINICAL

Clinical data has become the hallmark of evidence for a new medical product. Some level of experience in a human patient is typically required for regulatory purposes, to demonstrate support of marketing claims, to convince payers (private or governmental) to reimburse health care facilities for the treatment, for physicians to consider a change to their standard of care, and perhaps most important, for potential investors to assess the risk of investing in the product or company.

Clinical evidence can range from a review of published literature to a double-blind, randomized controlled trial with subject follow-up spanning many years. In the current regulatory and reimbursement environments, particularly for novel medical devices, the level of clinical evidence required is substantially higher than ever before.

Clinical Trial Design

The design of a clinical trial is a key element in the success or failure of the study outcome. Critical considerations for a trial design include: 1) a clearly defined intended use of the product—the claims that will be made about the product; 2) the population in which the device will be used; 3) the length of time during which the subject could experience an effect of the product; and 4) the anticipated difference between the outcome of the test and control groups. These parameters will determine the scope and size of the study that will be required.

The study should be designed to assess, as objectively as possible, whether the device achieved its intended purpose. In a study to support a U.S. regulatory submission, endpoints are typically defined to evaluate the safety and effectiveness of a therapy. These endpoints then are used to determine the size of the sample patient population required to statistically detect whether the study succeeded or failed. Endpoints must be concrete, easily measured, and directly related to the intended purpose. For example, an effectiveness endpoint might be complete closure of an atrial septal defect, as determined by transesophageal imaging. If a claim of stroke reduction is ultimately intended, however, the closure endpoint will not support the ultimate intended claim.

Clinical trials must be conducted in the specific population of patients for whom the product is intended, while minimizing potential confounding variables. The trial design should define specific patient selection criteria that address requirements for inclusion as well as exclusion. Subject enrollment is the most challenging aspect in conducting a clinical study. One challenge is balancing the need to enroll as many subjects as possible with the need to exclude subjects with medical contraindications or medication interactions that could bias the outcomes. Another challenge is that not all patients who are candidates for the clinical trial will want to participate or will participate to the trial's conclusion. Tactics to ensure adequate enrollment include setting up referral systems with other physician groups, approaching previous patients who had been seen at the health care facility, and using websites to notify potential subjects of the study. Tactics to retain subjects in the study include issuing follow-up reminders and in some cases offering incentives for subjects to return for follow-up visits.

In order to understand the ultimate effect of a treatment, some level of follow-up on each subject is required. In some cases, follow-up might be through the end of the treatment day; in other cases, follow-up might be months to years. The duration of follow-up required will be based on the expected effect of the treatment and the time period required to detect any adverse effects. For example, long-term implants will require follow-up for multiple years, whereas resorbable implants might only need to be followed for six months beyond the point at which the material has been resorbed. It is throughout this follow-up period that data will be collected to demonstrate that the benefits of the treatment outweigh any risks of the treatment—in other words, that the treatment achieves its intended effect with minimal safety concerns.

The last design element for a clinical study is determining the number of subjects that are required to be enrolled. For studies that require the highest levels of evidence, a statistically justified sample size is used. As a simple rule of thumb, where it is necessary to detect a very small difference in an outcome measure, the sample size required will be very large. The larger the difference between the outcomes of the test group and the control group, the smaller the sample size that will be required. Note, however, that even though it might be possible to justify a very small sample size statistically, it is unusual for regulatory agencies to accept sample sizes of fewer than fifty to one hundred subjects.

Each of these study design elements should be documented in a formal protocol that is signed and controlled. Each participating investigator must agree to abide by the study design as defined in the protocol. And each research subject must give proper consent.

Clinical Quality Controls—GCP

The clinical study must be conducted in accordance with good clinical practice (GCP), which represents principles to essentially ensure that the safety, health, and welfare of each subject is protected and that the data collected are valid and reliable. Numerous government regulations—in the United States as well as other countries around the world—have been established to address these requirements. Standards of clinical practice have been implemented to include an independent review of study documents and informed consent materials by institutional review boards (IRBs) or independent ethics committees (IECs) to ensure the protection of human subjects. Other practices, such as investigational site monitoring, verification of source documents to study data, controls over databases used to store clinical data, and methods to clean and lock the data set for analysis, are intended to safeguard the data validity.

Level of Evidence

On the continuum of clinical evidence, the prospective, randomized-controlled, multicenter trials demonstrate the highest level of proof. But there are lesser levels of clinical evidence that offer value as well, such a literature searches with meta-analyses, or feasibility trials with small sample sizes to gain insights into trends rather than being fully statistically justified. Consider the purpose for which the clinical evidence is required when determining the best approach for collecting clinical data. In the end, any process always should be documented in a formal protocol, and the objectives and endpoints must be clear and consistently applied.

My Advice:

Don't skimp on clinical evidence. Scientifically validated data can answer many questions from investors, customers, and potential acquirers.

CHAPTER 8

MANUFACTURING

Manufacturing is a very important topic, specifically when determining the estimated cost to produce the product. We need to understand the relationship between the cost of goods sold (COGS), the estimated average selling price (ASP), margins, and the current/projected reimbursement for the clinical procedure.

As an example, if a product has a projected average selling price, or ASP, of $1,000 to the clinic or hospital facility, and the COGS for this product is $900, then the gross margin is $100 (10 percent), which is an unacceptable profit margin. If the COGS cannot be resolved through improving manufacturing efficiencies to include reduced cost of materials, reduced direct labor costs or altering the design of the product, or if the projected ASP cannot be increased, then the product likely will not be successful as a business. The inventor should find this out early in the process. Understanding the target operating margins for diagnostic, therapeutic, and capital equipment is very important and will vary.

My Advice:
Make it for a buck, and sell it for four bucks. Understand the numbers early.

Manufacturing Competencies

The decision to use contract manufacturing verses the concept of self-build for your product is an important one for cost, investment, and control purposes. The advantages and disadvantages for each are much too complex to be addressed in this section. For the purposes of this book, we will focus on the contract manufacturing model.

The selection of a contract manufacturing partner is an important decision to assure alignment of technologies, customer service, financial expectations, and cultures consistent with the mission and goals of the inventor and entrepreneur. For example, the entrepreneur's product may require that the contract manufacturer provide key enablers, such as the technologies of injection molding, electronics integration, intricate assembly, and sterilizing. An assessment of contract manufacturers should be carried out to determine the best fit of these technologies as well as the candidates that have the technology foundation in-house or through a network of suppliers to meet the product needs. The selection criteria will be focused on technical capabilities but will also need to ensure that the vendor has the appropriate certifications and clean room space to manufacture the product. For most product applications, a minimum of International Organization for Standardization (ISO), ISO 9001 with Current Good Manufacturing Procedures (cGMP) will be required, but there likely will be a need for ISO 13485 and an FDA manufacturing assessment of the operation to make sure the operations are in compliance with FDA requirements.

Cost of Manufacturing

The concept of cost of goods sold (COGS) and its relationship to reimbursement and the end-user selling price to the customer all go hand in hand. Initial estimates should be considered early in the product development process, even though limited information may be available.

COGS:

➲ Cost of materials or components from vendors + direct labor + indirect labor + overhead related to manufacture of the product = COGS

Establishing the cost for a manufactured product requires close collaboration with vendors, whether the product is primarily manufactured in-house or primarily manufactured by a contract manufacturer. Most medical device entrepreneurs align themselves with a reputable contract manufacturer, rather than develop the in-house expertise.

Manufacturing a product in-house requires significant production and warehousing space. It also requires highly competent workers to develop the manufacturing process, purchase raw materials, hire and train employees, assemble the product, test, ensure quality compliance, sterilize, etc. Most young companies choose to contract with a manufacturer to produce their product, rather than make the significant investments in time and money to manufacture the product themselves.

The estimated purchase cost from a contract manufacturer for the entrepreneur's product will be dependent on volume and on the level of support that the contract manufacturer assumes is required. Additionally, there likely will be a number of one-time charges required to produce the product. These may include non-recurring engineering costs, such as tooling for plastic parts, packaging artwork, and initial design work for items such as electronic boards or circuitry.

Typically, 10 to 25 percent of the total contract manufacturing cost will be for engineering and support. The manufacturer's sales and administration costs could fall between 2 and 6 percent of the final cost. Finally, contract manufacturers look for margins in the range of 15 to 30 percent. These three factors, plus the cost of raw materials, direct labor, and overhead, comprise the total contract manufacturing cost to the entrepreneur.

Applying the logic described above allows the entrepreneur and the selected contract manufacturing entities to work together to understand the components and establish the estimated product cost range. The value in establishing this is to have an open dialogue with contract manufacturers to understand their commitment to the project and their openness to provide value commensurate with the expected reward. The vendor offering the lowest cost may not have what it takes to provide for an entrepreneur's needs and conversely, the highest cost does not assure that the vendor will deliver on its commitments. Finally, it is highly recommended that the entrepreneur seek references from others who have worked with the contract manufacturer. Work the network and find independent sources who can give an honest opinion of the contract manufacturer.

Over time, model refinements in the design of the product, operational efficiency improvements, and higher volumes will help to drive COGS down. As the product matures in its life cycle, the entrepreneur should assume selling price point erosion. The ongoing improvements in COGS will help to offset this erosion.

My Advice:
Tight margins in manufacturing will kill your company, period. Investors want to make money.

CHAPTER 9

FINANCE AND ACCOUNTING

When launching your company, establish accounting and financial controls early. Separate personal and business finances. Ensure you create an initial budget and track all flows of cash. The many reasons for maintaining solid financial control early in the process fall into two categories: defensive and offensive. The defensive reasons include ensuring compliance with federal, state, and local tax requirements; maintaining good order in case your bank requests access to financial statements; being prepared for investors requests of financial statements; and protection from potential fraud by employees, vendors, or other parties. A primary offensive reason for establishing accounting and financial controls is to use the information to manage and grow your business. Review your cash position, accounts payable, accounts receivable, revenue, and expenses, and ensure they are in line with your expectations. Manage your actual performance against projections and/or budgets.

Financial accounting and internal controls are important to establish at the inception of your company. It is important to have reliable financial information for management to run the business as well as for preparation of tax returns and dealings with lending institutions. A qualified accountant or CPA using accounting software should help you establish a monthly reporting cycle. In addition to the basic balance sheet, income statement, and statement of cash flows that every business should consider, an accountant can also help you identify the appropriate metrics for monitoring the key performance and critical success factors of your business plan. Establishing appropriate accounting procedures in the early stages of your business will provide a foundation for future growth. Reliable financial history can play a significant role in attracting future investors and venture capital resources.

My Advice:
Inventors usually focus their attention on the invention and its application and ignore the financial data. Master the financials of profit and loss, cash flows, and balance sheets—your future investors already have.

CHAPTER 10
CORPORATE STRUCTURING

Selecting the appropriate structure for your business can depend on your economic, tax, and governance needs. Consideration should be given not only to the appropriate structure in which to organize and operate your business but also to your exit strategy. Certain types of state, local, and federal tax credits may only be available for certain types of structures (including C corporations, limited liability companies (LLCs), and others). Regardless of the form chosen, the inventor needs to select a name for the company, register with the secretary of state, and seek legal assistance in creating the Articles of Incorporation, Operating Agreement, and Member Control Agreement, if investors are involved. It's important for you to work closely with your legal and tax team to develop organizational documents and agreements that support both your short- and long-term business strategies before you begin. Other important considerations will include where the company should be located (preferably in a state that offers incentive tax credits and research and development credits for start-up businesses).

My Advice:
Obtain professional advice. The money that you spend will be well worth the investment. Trust me.

CHAPTER 11

INTELLECTUAL PROTECTION

Patents

The topic of patents is very important to the successful inventor. Consult with a well-respected patent attorney, preferably one experienced in protecting inventions related to health care or medical technology and also experienced in working with start-up companies. In addition, selecting a patent attorney at a well-respected law firm can be helpful in assuring investors that your intellectual property is in good hands. Patent applications are not cheap; typically, you should be prepared to spend around $10,000 for a high-quality patent application.

There are important deadlines associated with patent filings. The invention should be kept secret until you have consulted with a patent attorney to be sure that you do not inadvertently forfeit your patent rights. As a general rule, a patent application should be filed before any non-confidential disclosure of your invention outside of the company. Failure to do so may jeopardize your patent rights.

In addition to protecting your own intellectual property, a patent attorney also can help you to evaluate the risk of your invention's infringing on the patent rights of others. Companies in the health care and medical technology industries are aggressive with patent filings and enforcement of their patent rights. A start-up company is well advised to consider these risks before investing significant time and resources into a new product. Be forewarned, however, that such an investigation can come with a significant price tag and also might reveal information that your company would rather not know. For these reasons, many emerging companies choose to move forward without such an investigation.

These simple tips can avoid misunderstandings with another party:

- ➲ Document your idea in a bound lab notebook, and have a witness sign and date each page of the invention description.
- ➲ Have an attorney that is familiar with intellectual property law review your idea and provide preliminary recommendations on whether to file a patent application. Provisional patent applications can sometimes be used to reduce initial costs but can impair future patent rights if not used properly. Prior art searches (i.e., investigation of past patents and publications in your subject matter) may be useful to evaluate the state of the art but are optional and will add additional expense.
- ➲ Prior to providing any information that may be confidential to anyone outside of your company, obtain a signed confidentiality agreement or non-disclosure agreement (NDA). There are many variations in NDAs, but all NDAs should include a statement that the recipient of the confidential information agrees not to disclose the invention to others, to ensure that the invention is not publicly disclosed by the recipient. Better yet, file the patent application before having any discussions outside of the company.
- ➲ Formally document in writing any of the confidential information that you have provided to the company, and prominently mark as confidential all documents or correspondence provided to the company.

If you do not obtain a signed NDA or file a patent application before having discussions outside of the company, you risk forfeiting some or all of your patent rights. This is especially true of patent rights outside of the United States, where a public disclosure can result in permanent forfeiture of patent rights in most foreign countries. Although the United States currently provides a one-year grace period to file a patent application, foreign countries are not so generous. Few companies will pay an inventor for an invention that the company cannot protect with a patent. Pending legislation in the United States, if passed, may further accelerate the need to file the patent application earlier rather than later.

Trademarks

Trademarks are generally not a high priority; the inventor should not be overly concerned with trademarks in the early stages of his or her company. Generally, inventors are interested in creating a company or product that will result in a "liquidity point," or exit, such as an acquisition by a large company or an initial public offering. Acquirers and investors usually are not impressed with excessive activities in trademarks, and they place little value in them at the early stage; in fact, trademarks can become a needless distraction of resources. Before significant resources are invested in a particular company or product name, however, it is a good idea to consult with a trademark attorney to ensure that the name you have selected will not infringe on the trademark rights of others.

Confidentiality Agreement

The inventor should have a standard confidentiality agreement obtained from legal counsel. Confidentiality agreements are also known by various other names, such as a non-disclosure agreement (NDA) or a confidential disclosure agreement (CDA). All confidentiality agreements are not alike; your attorney can provide you with an agreement suitable to meet your needs. The confidentiality agreement should be used for all discussions outside of your company, including discussions about the invention, strategy, marketing, operations, manufacturing, R&D, or other topics, at least until your patent applications have been filed or you have consulted with your patent attorney. If a person or company is unwilling to sign the confidentiality agreement, consult with your patent attorney before proceeding with further discussion. Venture capitalists and many large companies, however, usually will not sign a confidentiality agreement.

My Advice:
Seek high-quality legal advice. Never use online legal services.

CHAPTER 12

HEALTH CARE ECONOMICS

In today's cost-sensitive health care climate, technology companies can no longer assume that the market will bear increased costs to obtain improved clinical outcomes. Even with a compelling clinical story, technologies will not be adopted into use unless the return on investment (ROI) can be clearly demonstrated.

Health care economics should be considered at the time of initial development of a product—an important component of raising capital. Later, while preparing for market introduction, technology companies must consider the financial advantages and disadvantages for each economic stakeholder and tailor messaging according to their unique needs. It is advisable to avoid strategies that deliver benefits to one stakeholder at the expense of another. For example, relying on CMS and/or commercial payers to separately reimburse the incremental costs of a new therapy is a risky approach. Even in the rare case where this is successful, payer policies and contracts are frequently revised, and payers are unlikely to bear the increased expenditure for long unless they recoup the savings elsewhere. The inventor should begin early in the business process of developing a clinical-economic model that demonstrates how the technology fits into the overall health care system. Ideally, the new medical technology will support a financially compelling story for every economic stakeholder.

Economic Stakeholders

Centers for Medicare and Medicaid Services (CMS) and commercial payers are interested in reducing overall costs associated with the care of their subscribers. They will consider expenditures across the continuum of care, typically within an approximately eighteen-month window (as longer-term economic gains may not be realized by the payer responsible for the initial investment). Payers are increasingly avoiding financial responsibility for rising technology costs by devising "bundled" or "packaged" payment methodologies and eliminating carve-outs or distinct payments for products and services. This places the onus on hospitals/physicians to demand value from their technology investments.

Hospital administrators tend to view all new technologies as incremental costs and therefore a threat to their increasingly tight margins. Before a purchase is made, they will demand information on how the financial investment might be recouped (e.g., through efficiency gains and/or reductions in other material costs). Conversely, hospital administrators are not likely to be motivated by data that demonstrate a reduction in the need for clinical services (e.g., less follow-up care or a reduction in the reoperation rate), as this could actually translate into reduced reimbursement and/or revenue opportunities for the facility.

Physicians are increasingly likely to be aligned with the hospital administration's focus on minimizing expenditures, as they may be employees or owners of the hospitals where they practice. In addition to their growing desire to maintain cost-effectiveness for the facility, physicians will be interested in the impact of a new therapy on their professional compensation.

Fueled by the national debate on health care reform and the rise in health savings-account plans, patients are increasingly motivated to control their health care expenditures. They are proactively requesting details regarding the cost of their care and demanding transparency in hospital and physician charges. Patients want to know whether they will bear responsibility for the cost of new technologies, either directly or indirectly, in the form of increased premiums.

Addressing the varied needs of these diverse stakeholders can be a challenge, particularly in a rapidly changing health care policy environment. Formulation of an economic strategy should be initiated early in the development process, parallel to the clinical strategy. A professional with health care economics expertise can provide substantial support by guiding companies to identify appropriate clinical variables for collection and analysis (i.e., those that can be quantified economically), establishing appropriate and defensible pricing for the new technology, and designing interactive economic models for use in communicating a compelling value proposition.

My Advice:

Your ability to access the market will be dependent upon your ability to support the return on investment for each economic stakeholder. Anyone with an economic stake in the adoption of your technology is a customer with needs that must be addressed.

CHAPTER 13

SELECTING THE RIGHT TEAM

The selection of the team of people to assist in creating your product idea is probably the single most important thing that you will do. First, you should have a general consultant who will lead the project for you. Generally, this is a person with strategic marketing skills who has the ability to ask all of the right questions and has most of the right answers. It should be a seasoned and accomplished professional who can see the bigger picture, who will lead the multifunctional team, and who will be accountable to you, the inventor. You will need to rely on and trust this person and his or her sound advice and guidance as he or she leads a team of experts in their various areas of expertise.

It is paramount to remain objective during the discussions and to review data, including market research and other information. Many inventors and their opinions are often biased; they are too close to the product idea and will hear what they want to hear versus what they need to hear. Sometimes inventors will seek numerous opinions from others who are not qualified to provide advice and will become confused. Avoid this pitfall.

The chart on page 40 (Figure 1) outlines the major areas of expertise and functional areas that will be needed on a limited basis to develop and prepare your medical technology as it moves through development on its way toward a market launch and the fulfillment of satisfying proof of concept and proof of principle. This illustration is identical to how medical technology companies organize their project staffs for the commercialization of their projects.

My Advice:
Hire consultants who are accomplished and experienced and who can communicate effectively. Surround yourself with the best talent. Remember, talent is everything.

Inventor's Consultants—Virtual Staff

You will need to follow a commercialization process that requires that people with very specific skills serve you, under the direction of a centralized contracting service-provider consultant CEO, or a hired employee CEO at the appropriate time. There are stages in your company's growth where both will serve you well. In the earlier stages, a consultant usually will meet the need because of limited funding and the requirement of only part-time and limited involvement. As the company grows, however, and additional funding is secured and the intensity of milestone achievement requirements increases, investors and you, the inventor, should move to the employee model for a CEO of the company.

The inventor CEO is an important topic and frequently a sensitive one as well. Often, inventors feel obligated to be the CEO and key influencer of all activities. He or she will frequently attempt to act as his/her own general contractor by providing direction and making all decisions on the operation of the commercialization process, many times by directly contacting each service provider in separate "silos," without the coordination and collaboration that is required with all contractors of the staff. This is a critical mistake. Seeking out the right people for each of the identified areas and having them work well together, as shown in Figure 1 on page 40, is very important to the success of your invention.

My Advice:

The virtual staff is a contracted staff of professionals who are experts in their respective fields. They are best managed under the control of one consultant/general manager or CEO who is accountable to the inventor and owner (you). Managing this process and all of the skilled people on your own is one of the biggest mistakes that an inventor can make. The number one reason for a failed start-up is poor execution.

Virtual Staff

Figure 1 Consultant Cross-Functional Staff

While there is a natural tendency to want to form a company, lease office space, and hire a team of people, it is far more desirable in the early stages to have a virtual company staff. Consultants provide the most common pathway to follow in the early-stage start-up for reasons of cost savings and high skill expertise. Investors will want to know and understand with whom the inventor has teamed up to create and execute the business plan and the business. The virtual staff of contractors conserves on the expenditures of financial resources, and their work activities can be adjusted up or down to meet the needs of the organization. (See Figure 1 on page 40.)

My Advice:
You would not dream of building a luxury home by yourself; your company is a similar situation. You need a general contract consultant who works well with the other contractor staff experts.

Employees

In the early stages of a medical technology company, some inventors consider hiring people to develop their product. This decision should be handled very carefully, as the cost of hiring employees can drive up the cash-use rate, or as it is more commonly referred to, the "cash-burn" rate. Investors do not look favorably on the prospect of hiring employees prematurely but will usually favor the use of consultants. The advantage of consultants is that their time and cost can be more easily controlled. Additionally, by adding employees you create additional obligations and legal exposure to your company. Finally, if an entity is interested in acquiring your company, it generally is far more interested in your technology. Employees and "brick-and-mortar" obligations are often of little interest and can be considered an obstacle during the acquisition phase.

My Advice:
Do not hire employees prematurely. Investors do not like to see financial resources spent this way. The virtual contractor/consultant staff is the preferred structure.

CHAPTER 14

YOUR ROLE IN FUNDING

The inventor's role is critical in the start up of the company. Financing the invention initially will likely be the responsibility of the inventor, by use of his or her private funds and sweat equity. The next source of funding will include what many refer to as the three F's: founder, family, and friends. Other sources include loans and credit lines from a bank.

Initial funding by you, the inventor, and the three F's is to fund the "investment story." Your initial investment has one object: to fund the activities that will build your story to get you the money from angel investors and venture capitalists. These activities include but are not limited to initial market research and due diligence plans for validation and verification of the business; the business plan and financial schedules; valuation and share structure determination; investor presentation; legal support activities for raising funding; and retained activities to assist you in the introductions and presentations to potential investors.

The next major step in funding includes private funding sources of other investors, which include angel investors and venture capitalists. Some inventors confuse these groups by thinking that they are the same; they are not. Angel investors usually fund earlier in the development cycle with smaller amounts of funds, and venture capitalists follow on later with larger sums of funding. Another difference to remember is that angel investors are investing their own private individual funds, while venture capital people are fund managers who generally invest other people's money, not their own. The traditional stereotype of the gray-haired and benevolent angel investor who was wealthy enough to write a check for an interesting idea is generally no longer true. Many angel investors are organized groups of well-to-do and sophisticated investors who may require nearly the same level of investment due diligence rigor as venture capitalists. Because venture capitalists are professional investors who use other people's money, they are sophisticated business and financial people who will require that the inventor go through a very thorough process of evaluation and will have significant demands for their rights for a certain level of control.

Angel investors and venture fund managers will require the inventor to have some "financial skin" in the game as the initial investor. The investor wants to know that the inventor is committed and is passionate about the invention and that he or she is in it for the long haul. The arrangement is analogous to building or purchasing a home. The lender requires that the purchaser contribute a down payment of between 10 and 20 percent. Angel and venture capital investors are no different.

The amount of investment expected from the inventor is generally proportional to the financial wherewithal of the inventor. Take the example of the graduate engineering student who has a promising idea but needs to raise a $500,000 angel round. Investors will expect some level of investment by the student, but it may be proportional to his or her ability to raise funds from his/her own resources and family and friends.

On the other hand, if it is a physician inventor who owns a stand-alone surgical center that is well established and is very profitable, a higher initial investment likely will be expected from the physician inventor. The inventor should outline and understand the activities that he or she will be required to fund. Failure to understand this is a common mistake made by inventors. The responsibility for this initial funding can include the funding of the patent filing, market research, business plan and the preparation, and activities for raising additional funding. It could also include the possibility of developing early prototypes for proof-of-concept activity in animal models.

Investors are interested in learning the answers to certain questions that include:

- ➲ How sure is management that the amount is sufficient to successfully execute key milestones?
- ➲ What is the payback or return on investment that will be realized?
- ➲ How much capital is needed?
- ➲ When will it be needed?
- ➲ How will the investment resources be used?

An experienced team can help the business quantify these factors in a business plan and financial schedules. A quality business plan will help with the determination of potential buyers and create expectations for all parties involved. Brilliant ideas do not always translate to business success. A well-formulated and executed business plan will enable management to maintain control of the business and achieve their goals.

In conjunction with the business plan, the financial projection is one of the key tools of a start-up business. The financial projection has an impact on both the funding of the business and the operational aspects of the business; it demonstrates the thoroughness of the business plan, and it is a means for you to translate your great idea or product into the common denominator of dollars and cents. It is the "instrument panel," or "dashboard" of your business.

The preparation of the financial projection is usually a highly collaborative effort that involves the inventor and entrepreneur, along with finance, marketing, R&D, manufacturing, and operational functional leaders of the team and other professionals. From an operational standpoint, the financial projection provides a budget or measuring stick against which to evaluate the financial progress. Use of the projection as a control to measure milestone progress against the actual results will help management focus in on changes that may need to be made to the business itself over time.

From a funding standpoint, the financial projection provides guidance to how much funding the business will require for development, working capital, and the eventual selling of its product. It also will provide data for investors to evaluate their potential return and weigh your business plan against the others they are considering.

When first-time inventors and entrepreneurs consider and begin to document the proof of principle, they usually start to sketch out a rough idea of how much it will cost to bring the product to market, how the product will be sold and for how much, the size of the market, and how quickly the product will obtain a certain market share. These ideas are the beginning of a financial projection. The projection will also consider how to staff the business, how much inventory the business will carry and when, what terms to provide to customers, what terms to provide to vendors, how much debt to carry, and how much capital to raise and under what terms.

Often, the first thought associated with the projection is the income statement, but the income statement is only part of the equation. Using the expectations for inventory, accounts receivable and payable, debit, and equity discussed previously, a sound projection will include the business balance sheet, income statement, and cash flow statement, as well as fully documented underlying assumptions.

My Advice:

You need to have "financial skin" in the game. Investors are not going to carry 100 percent of the financing. They want to know that you believe in your product and business and that you will not walk away from the deal.

CHAPTER 15
ANGEL AND VENTURE CAPITAL

There are numerous funding-source options for your business. It's important for you to develop criteria to find the most cost-effective funding source, including various debt and equity considerations. Be sure to weigh the benefits and challenges with the funding alternatives. Angel investors can help to establish the viability of your business, which is usually followed by venture capitalists, once certain benchmarks have been met and the need for capital accelerates. Competent legal and tax advisors can help you develop agreements with your financial partners to maximize your economic, strategic, and tax goals. Your legal and tax team will work with you to establish the appropriate incentive mechanisms to align all parties' interests, as well as contemplate exit strategies, based on your longer-term goals.

The important thing to remember when presenting to investors is to have a highly effective story in the executive summary, business plan, financials, and slide presentation, as required by the investment group, and to keep in mind that you are not selling a product; you are selling an investment.

My Advice:
Terms to know and understand include seed, pre-angel round, angel round, and Series A, B, and C financing. Your general consultant can help you understand the relationship of these financing steps.

CHAPTER 16

DEAL STRUCTURE CONSIDERATIONS

You have spent countless hours of precious time and resources planning the development of your innovation. The time has come where your resources aren't enough to proceed, and you need to seek outside capital. Knowing how to fund the deal is perhaps the least understood and weakest point of preparation for most inventors and entrepreneurs. This is painfully ironic, considering it is likely the most important aspect of your development cycle—no funding, no commercialization. The heartening news is that with proper planning and implementing a developed strategy, most technologies will receive funding (if they follow the development cycle in this book).

It is critical to note that the objectives of the entrepreneur and the investor are often juxtaposed with one another. As a result, this is an important area to understand very early in the start up of your company. The investor sees the best deal structure as one that provides common incentives to both parties and, ultimately, a successful exit from the company. The entrepreneur typically wants to retain the highest percentage of ownership (and therefore, control) of the company for himself or herself as will the investor. Repeat or serial entrepreneurs more readily grasp the need to avail a portion of the company to outside investors, not just for necessary funding but also to open the company to strategic investors who can attract future capital or industry access that will be essential for the early-stage company.

When negotiating the financial terms of the deal, consider three basic areas: First, ensure that the parties in the operation of the business—founders, investors, team members—are aligned, including ownership, capital structure, strategic vision, and team-building strategy.

Second, it is imperative to recognize the role of the board and how to build a strategic yet cohesive team of investors and industry experts to guide the decision-making process of the companies. Companies should expect the board to be involved in creating, enacting, and validating the strategy of the company. The days of passive board membership and infrequent communication with the executive teams are over. Highly effective early-stage companies often have highly effective boards.

Third, consider the value generation of the investment. Investors do not put money into an early-stage company and then allow it to be steamrolled by subsequent rounds of investment. The terms that are negotiated, often with great parley, bring the investor and the founder into alignment, and the investor needs assurances that the company and its team will deploy the capital responsibly to build value in the company.

These are all complex areas that need to be defined clearly in the terms of financing the deal so that disputes will not break out later and threaten the company's success. These topics are significant in nature and require the attention of professionals in the practice of corporate law and accounting, with specific experience in the medical technology start-up world.

The important thing to remember is that both the entrepreneur and the investor must have commonly aligned goals for success and that the two parties should negotiate to a point of mutual satisfaction.

Private Placement Funding

Funding through angel investors and venture capitalists must be compliant with state and federal governments' rules for the issuance of securities. Most early-stage companies obtain funding by implementing a private offering, which is allowed in some form in each state for intrastate offerings and under U.S. federal law for exempt offerings (offerings that are not publicly registered and offered for purchase to the public). The sale of securities is highly regulated; depending on the amount of capital necessary for your company and when you need it, great care should be taken to determine if an exempt offering (an offering exempt from public registration) is advisable under a state or federal exemption. State law and the Securities Acts of 1933 and 1934 are the regulating law for the sale of securities. Navigating the nuances of these sometimes illogical rules will make seasoned legal counsel a welcome resource. Most private placement is offered under rules known as Regulation D. Please note: an offering of the sale of securities in more than one state is covered by federal securities laws. Most private offerings today elect to be exempt under a federal exempt offering, due to ease of use and federal preemption, which allows a nearly uniform offering to potential investors in various states. An offering that is exempt under federal rules *does not* exclude a company from following state rules for an offering or an exempt filing. These "blue sky" considerations must be considered in advance of structuring an offering or considering investments from investors in certain states to prevent regulatory scrutiny, which could jeopardize your entire offering.

An exempt offering usually is directed toward a small number of select private investors. Under federal rules, these investors are individuals or entities who have a prior relationship with you, as you are largely prohibited from any form of solicitation to potential investors whom you don't know or with whom you lack a relationship. In recent years, securities offered to investors are usually with additional rights and of a preferred nature to the founders and common shareholders, who are usually angels. The terms of an offering are typically put in a term sheet that often follows the form established by the National Venture Capital Association and embedded in a "private placement memorandum" (PPM), which describes the terms of an offering for the sale of securities being offered to the investor, with all of the business terms, financial projections, risk factors, business strategy, and legal disclosures. Offering memorandums are not for the faint of heart and can be a time-consuming process. That being said, the offering memorandum is the salient document that takes your business plan, financial projections, and deal terms into one form; it is truly a watershed document. Nowhere else will an entrepreneur need to think about all aspects and phases of the past, present, and future of the company in one document for potential investors to see. It is in the PPM that all of the questions pertaining to regulatory requirements, reimbursement, clinical data and trials, market adoption, strategy, competitive factors, financial projections, and legal risk are wound together into one cohesive document.

My Advice:
Seek professional corporate legal help in the medical technology space early in the start up of your company. Begin to think about the structure and exit at the beginning.

CHAPTER 17

LAUNCHING IN THE UNITED STATES

The United States is the largest health care market in the world, with the highest average selling price for medical technology products. The U.S. health care system remains, for the most part, a fee-for-service payment system for providers, with significant opportunities for the medical technology start-up, including opportunities in Europe, Asia, Latin America, and other parts of the world. If you are an inventor from these regions, several additional important considerations must be understood, including U.S. customs rules and regulations, for the importation of your product. Your general contractor consultant should assist you in understanding these regulations and assist you in the paperwork that satisfies customs requirements. Medical technology products that are imported into the United States must be FDA-cleared or approved, generally, but there are exceptions to the rules in non-selling or marketing situations. Check with your general contractor and the U.S. Customs Office for further details.

Conducting an appropriate level of due diligence and market research is very important before attempting to commercialize and launch your product in the United States. Cultural, reimbursement, regulatory, and provider practice norms may differ significantly from the inventor's homeland in Europe, Asia, Latin America, or other parts of the world.

My Advice:

Understand the U.S. market opportunities and obstacles and the rules for importation to the United States. Verify and validate the value proposition in the U.S. market early in the process.

CHAPTER 18

THE EXIT

The exit is something that should be considered—and considered seriously—at the start of your company and revisited throughout the process of developing your product and company. You need to consider the realistic options for an exit but realize that you have little control over this event. What we are talking about is the decision to build to run your company vs. build to sell it. I always believe that you build to run your company because you have little control over when or if someone will buy your company. Remaining focused on performing the right tasks in building to run the company vs. building to sell the company is the wise approach. Inventors sometimes make the future sale of their company the priority and miss the key steps in building it in the proper way.

So you will need to rely on good advice and experienced people in helping you with this important topic. As the inventor you may have already decided that you have no intention of building a brick-and-mortar company or even fully commercializing your product. Maybe you simply want to take it far enough in the development cycle for a potential acquisition or for a strategic investment by a larger company that may ultimately acquire the product or company. You have to remain focused on the build-to-run approach for the company while working toward an exit. In most acquisitions situations, the acquiring company is not interested in bricks and mortar, employees, or even long-term supply agreements that inhibit its ability to assimilate the product/technology into its existing infrastructure.

Initial public offerings and strategic buyers are an exit pathway for the inventor, one that both angel and venture capitalists are very interested in as well. Angel and venture capitalists need to move their investments to a more mature state so that they can either be acquired by a larger company or be offered in the public markets as an initial public offering, thus creating a profit and freeing up their cash for their next investment.

As a rule, planning your exit and preparing your company for your exit should happen years before you expect it to happen. Consider the following things to help you better prepare:

⮞ Build a strong management team. Companies that have a strong management team sell for more than those that rely on the talents of an individual (or individuals). Investors, especially venture capitalists, invest in management teams as much, if not more, than in technologies. When faced with choosing either an excellent technology and an average management team, or an excellent management team and an average technology, most would choose the latter.

⮞ Manage your financial statements to report the greatest income and cash flow in the years leading up to your exit, and stop managing the net income so that the least amount of income tax is paid. Example: Reporting an additional $1,000 of expense may save you $400 of income tax, but it could cost you $6,000 or more in value—hardly a fair trade.

⮞ Have your financial statements audited, if they are not already. Audited financial statement provides potential buyers with the highest level of confidence in the company's financial condition and performance, which nearly every time leads to a higher sale price.

- ⮑ Seek competent tax advice. The structure of how the deal is put together can have an enormous impact on taxes that need to be paid. There are stock sales that are treated as asset sales, and asset sales that are treated like stock sales.

- ⮑ Find a competent merger and acquisition (M&A) advisor to assist you through the selling of your company—never try to do it on your own. A knowledgeable professional also will know the market and who the potential buyers are. He or she will be able to market the company in a way that will attract the greatest number of potential buyers and create an atmosphere of competitive bidding.

- ⮑ Strategic buyers usually will require that proof of concept and proof of principle, as well as clinical data, are demonstrated to their satisfaction. They also usually acquire the best in class in a technology class, and the technology must be in an area that fits their needs and goals.

- ⮑ Choosing the time and method for an exit is a business decision, not a legal or accounting decision. Be clear as to why you want to exit and then choose the method that best suits your goals and those of your Board of Directors, always keeping the welfare of your investors as your highest priority. They got you to where you wanted to go, and now they deserve their reward, just as you do.

- ⮑ Consider all options after consulting with your Board of Directors members and legal/accounting team.

My Advice:
Start planning for the exit as you start the business. Good legal, accounting, and M&A professional advice will support this goal.

CHAPTER 19

EXECUTION, SUCCESS, OR FAILURE

The success of your medical technology product will be heavily dependent on the execution of your business plan, but remember that the business plan will likely change many times between the first draft and commercialization. Adaptability and evolution are extremely important. Most start-up companies and products fail because of execution-related issues. This is why angel and venture capital people carefully evaluate the management team, which includes the consulting team that you have chosen.

Success

Success comes down to who you've chosen to help you, along with performance of *the right tasks, in the right order, in the right way, by the right people, with the right skills at the right time*. Sound complicated? It's not, really—if you have the right people to get you started. Many inventors choose the wrong people to help them get started. An engineer inventor may hire his friend, who practices family law, to file his patent; a physician inventor may ask his favorite sales representative to be his CEO. This type of decision making can be the kiss of death to a start-up. I think you get the idea. Seek out the right people with the right skills at the start. The most important type of person to seek out is one who will act as the project lead or general consultant for your product or company. This is especially important in cases that involve physician inventors or other people who have very limited time to dedicate to the project and lack the experience that is needed for success.

Failure

Remember, if you fail, fail fast and fail early—early in the development process, not at execution of the launch. It is less painful, financially and emotionally. I know that this is a tough topic, but it is better to determine very early that the product will not be commercially successful than to invest hundreds of thousands or millions of dollars, only to fail many months or years later. The road of medical technology is littered with attempts by inexperienced inventors who failed with their initially promising technology in a questionable market segment, many times because they did not validate the business early enough, or they selected the wrong people to plan and execute the enterprise. The important point missed was that the inventor did not validate and verify the market opportunity effectively with an initial quality assessment. Do your homework early, before you spend a lot of money and hard work on developing something for which a market does not exist. And select the right people to help you. You will be glad that you did.

My Advice:

It is all about execution. Most companies that fail will fail in executing. The management team is paramount. Remember that the first investment in your company should be to validate and verify through market research, surveys, and focus-group activities.

CHAPTER 20

WHAT IS THE INVENTOR'S ROLE?

Your role as the inventor and owner of your business is very important. You must be involved in the operation of your company—but there is a delicate balance between being involved and feeling obligated to make every decision. That is why you have the best talent around you to increase your chances for success.

It will be very important to make yourself available to the contractors that you engage. Initial face-to-face meetings are ideal, and then availability on a daily basis via e-mail, text, or phone is very important as the project unfolds. Frequently, inventors are not easily reached once work begins, and this leaves the contractor without the needed information and lacking important input on decisions that need to be made.

Your Responsibilities
- Be reasonably available to the general consultant and the consulting team.
- Attend weekly or other periodically scheduled staff meetings as requested via phone.
- Respond in a timely fashion via communications channels such as e-mail (within twenty-four hours, maximum), phone calls, and even text messaging.

- ➲ Remember that the consultants you have selected are acting in your best interest.
- ➲ Learn to understand things outside of your comfort zone.
- ➲ Trust the management team that you have selected—trust but verify.

My Advice:

As a start-up, you have taken on, at minimum, an additional part-time job. Be available to the consulting team, learn as much as you can, and seek mentorship from those professionals with whom you have contracted, especially the general consulting leader of the team.

CHAPTER 21

NOW THAT YOU HAVE READ THIS BOOK

It is very important for the inventor to follow the steps in the next paragraph to enhance the probability of success. Inventors commonly miss these steps or perform the wrong tasks in the wrong order without the assistance of skilled people, thus draining them of financial resources and leading to a premature failure of the product and company. Managing the inventor's expectations to complete the activities is very important. You need to invest in the activities that will be valued by future investors, not what you, as the inventor, want to do or think should be of interest.

To increase your probability of being successful, remember that this is a game of performing the right tasks, in the right order, at the right time, by the right people, with the right skills.

Step Summary for the Medical Technology Start-Up

The following is a list of the recommended general steps that you should follow in the creation of your start-up company. Approach each step in "General Success Steps" with an appropriate amount of planning, executing, evaluating of results, and taking of corrective action; then decide whether to proceed to the next step or to stop the project. This gated approach will increase the probability of success and manage the risk of your investment as well. Follow the pathway, and you will increase your probability of success; take short-cuts and skip steps, and you could be doomed to fail.

General Success Steps:

1. Your very first consideration should be to answer to the following question: What problem do you think your invention or idea will solve? Then determine the market opportunity. Complete the founders, friends, and family funding. *See chapters 2 and 14.* Secure your patent rights by consulting with a reputable patent attorney with medical technology experience. *See chapter 11.*

2. Identify an accomplished and experienced general consultant first, and seek his or her assistance in identifying a team of consultants in the needed areas. The general consultant should be a strategic marketer and business development person. This person is your "quarterback" on the staff—under your control. *See chapter 13 and Figure 1.*

3. Perform early due diligence, including market research, regulatory pathway, and reimbursement as initial estimates, to validate and verify the business proposition before you go too far and invest much financially. Ask yourself the following question: Do I potentially have a business here? Create an initial budget and timeline—and understand what it will take to be successful. *See chapters 1–9.*

4. Contract the functional staff of expert consultants with the assistance of the general consultant. *See chapter 13, Figure 1.*

5. Have a high-quality business plan with financial schedules written to include profit and loss (P&L), waterfall cash flow, and balance sheet statements. Update budget estimates to meet the projected milestones. *See chapter 14.*

6. Establish the company valuation estimates and share structure. Create the appropriate fund-raising documents to include a private placement memorandum (PPM), if required. *See chapters 15–16.*

7. Identify the appropriate angel investment group that mutually fits the needs of the inventor and the angel investment group. *See chapter 15.*

8. Prepare and deliver a succinct but high-impact investor presentation to qualified angel investor groups, and secure funding to targeted levels. Remember, you are presenting an investment opportunity, not selling investors a product. *See chapters 15–16.*

9. Draft and complete the investment agreement to close the angel funding round. *See chapter 16.*

10. Complete milestone requirements as presented to angel investors for proof of concept and proof of principle; prepare for potential early exit with your team. *See chapters 6, 18.*

11. Seek venture capital or private equity funding, as needed; understand the deal; re-evaluate share structure. *See chapter 16.*

12. Launch the product, build a sustainable company, or seek and secure an acquirer or initial public offering. *See chapter 18.*

My Advice:
Follow the steps. Deviation is death in the start-up world.

SUMMARY

Many inventors from medicine and the engineering fields have made countless contributions to the advancement of the diagnosis and treatment of disease, including Drs. Michael DeBakey, Kurt Amplatz, Julio Palmaz, Richard Schatz, Jeremy Swan and William Ganz, Thomas Fogarty, and Andrew Cragg, to name just a few. Never underestimate the impact that your medical technology idea can offer; never cut yourself short. It does take courage to step up and take the risk to make a difference.

Starting and managing your own medical technology company is much like being your own general contractor in the home-building business. Unless you're already in the home-building business or have significant knowledge and skill in the construction business, being your own general consultant is very risky, because you do not even know the right questions to ask. Your strengths are being an expert at practicing medicine, engineering things, or being a business-school graduate or student; or maybe you are a lay person with a creative mind. You are not an expert in all of the functional areas of developing a medical technology, and you probably have a day job and a family to boot.

Where should I start now that I have read this book?

The most important take-away point of this book is to *not* attempt to do these things on your own. Seek highly skilled and experienced people who can help you. Your next moves with your medical technology start-up company should be identifying an expert general consultant who can help you and following the "General Success Steps," as outlined in chapter 21.

Best wishes for success.

Patrick Kullmann, MBA
Founder and Lead Strategist, CG3 Consulting LLC

If you are interested in obtaining additional information, including access to our library of *free planning tools* for the medical technology inventor, contact me at *pkullmann@CG3Consulting.com* or at our Global Call Center, 1-866-CG3-8111 (243-8111). From outside the U.S., call 952-921-5851.

My Final Advice:
Seek and listen to the experts. You will sleep better at night.

AFTERWORD

One of my early business mentors used to say, "There's the smartest guy in the room, and there's the richest guy in the room, so which one are you?" I think this gets at Patrick's introductory point about the difference between an inventor (the smartest guy in the room) and an entrepreneur (the richest guy in the room). Believe me; it is extraordinarily difficult to be both—or at least both things at once. Knowing what you don't know is my number-one piece of advice for both inventors and entrepreneurs, and Patrick has done a nice job of pointing out the myriad steps and pitfalls to hammer at that point. Even recognizing this is a high-level summary of the pathway from concept to market for medical technologies, you should now realize there are a number of complex, interdependent, and risk-laden steps. Particularly as physicians, we should not delude ourselves into thinking that because we are a great at being an anesthesiologist, surgeon, cardiologist, etc., that by extension, we are also great at business. Just like learning a complex procedure in medicine, we need to have a good game plan and the right people to coach and mentor us. I echo Patrick's advice to find an experienced and thoughtful general business consultant as a first step in the entrepreneurial journey.

My second pearl of business wisdom: your business plan gets you a management team, and then your management team gets you the investment money. Having a good business plan is necessary but not sufficient; smart angel investors browse the business plan but closely scrutinize the experience and track record of your management team. Here again, having the right general business consultant can help you develop a solid business plan and then help you recruit and hire the right management team.

Best of luck to all the readers of this excellent guide, and special thanks to Patrick for creating it.

Timothy Hannon, MD, MBA
President & CEO, Strategic Healthcare Group LLC
Indianapolis, IN

ABOUT THE AUTHOR

Patrick Kullmann, with over 28 years of medical technology experience, founded CG3 Consulting LLC, a full-service commercialization advisory firm. CG3 represents creating growth in the health care, scientific, and technology industries. Patrick established the firm in Minneapolis and has since expanded to both the East and West Coasts for global coverage with offices in Boston and San Diego.

Prior to establishing CG3 Consulting, Patrick was a senior director at Medtronic in their $2.3 billion cardiovascular division. During his career he has served in ten medical specialties as a senior marketing, business development, and sales executive.

Patrick is a well-respected entrepreneur, speaker, educator, author, board member, and angel investor serving the medical technology fields. He has authored several articles for leading medical technology and other life science organizations in the United States and abroad. He has served in senior marketing, business development, and sales leadership positions at Medtronic, Boston Scientific, Baxter International, Johnson & Johnson, and four start-up medical technology companies, two of which were acquired for a total of $120 million.

Author's Publications and Speaking List

Publications authored by Patrick Kullmann

- "Rule of the Raise." *LifeScience Alley/World Medical Device Organization* newsletters, October 2010.
- "First Things First." *LifeScience Alley/World Medical Device Organization* newsletters, November 2010.
- "Valuation Principles for Early Stage Companies." *LifeScience Alley/World Medical Device Organization* newsletters, January 2011.
- "Is It a Science Experiment or a Product?" *LifeScience Alley* newsletter, March 2011.
- "Importing Medical Technology to the U.S. Market." *LifeScience Alley* newsletter, May 2011.
- Presentations by Patrick Kullmann
- "Capitalizing the Life Sciences Sector," Red River Valley Research Corridor—Life Science Summit Topic: April 18, 2011.
- "Entrepreneurism and Medical Technology," St. Mary's University, Minneapolis, MN.

Author's Professional Affiliations

- Society of Physician Entrepreneurs SoPE, Denver, CO
- LifeScience Alley, Minneapolis, MN
- BioCom, San Diego, CA
- MassMEDIC, Boston, MA
- BioForward, Madison, WI
- Red River Research Corridor, Fargo, ND
- Mentor Minnesota Angel Network, Minneapolis
- Mentor Minnesota Technology Cup, Bio-Division, Minneapolis–St. Paul, MN
- Accredited Angel Investor

INDEX